Healing with Data: Machine Learning in Healthcare

Er. MEHAKPREET SINGH

Er. PRIYA

Healing with Data: Machine Learning in Healthcare

Science | Technology | Machine Learning | Healthcare

Contact : officialmehakpreetsingh@gmail.com

Copyright © 2024 Authors

All rights reserved.

ISBN: 9798325631306

Preface

In recent years, the convergence of machine learning technologies and healthcare has precipitated a paradigm shift in the diagnosis, treatment, and management of diseases. As researchers and practitioners, we find ourselves at the forefront of a transformative era, where data-driven approaches hold the promise of revolutionizing healthcare delivery and patient outcomes.

The genesis of this book, "Healing with Data: Machine Learning in Healthcare," arises from a profound acknowledgment of the critical intersection between data science and medicine. The exponential growth of healthcare data, coupled with the increasing complexity of medical decision-making, underscores the pressing need for robust computational methodologies to extract actionable insights from vast and heterogeneous datasets.

Recent research data unequivocally accentuates the imperative for leveraging machine learning techniques in healthcare. Studies indicate that medical errors, misdiagnoses, and treatment inefficiencies contribute significantly to adverse patient outcomes and substantial economic burdens. However, with the advent of sophisticated algorithms and computational models, there exists a transformative potential to enhance diagnostic accuracy, personalize treatment regimens, and optimize resource allocation within healthcare systems.

Through this book, our aim is twofold: to demystify the principles of machine learning for healthcare professionals and to elucidate the manifold applications of these methodologies across various domains of medical practice. By fostering a nuanced understanding of machine learning algorithms, from foundational concepts to advanced techniques such as deep learning, we endeavor to empower readers with the knowledge and skills requisite to harness the full potential of data-driven healthcare.

Moreover, in cognizance of the ethical implications inherent in the deployment of artificial intelligence (AI) systems within clinical settings, we devote substantial attention to exploring ethical considerations, regulatory frameworks, and strategies for mitigating algorithmic biases. It is incumbent upon us, as stewards of technological innovation, to ensure that the adoption of machine learning in healthcare is imbued with principles of fairness, transparency, and accountability.

We invite readers from diverse backgrounds – clinicians, researchers, data scientists, policymakers, and students – to embark on a journey of discovery and enlightenment through the pages of this book. As we navigate the intricate landscape of machine learning in healthcare, let us collectively endeavor to realize a future where data-driven insights converge harmoniously with compassionate care, thereby ushering in an era of healing with data.

Mehakpreet Singh

Table of Content

Preface III

1. Introduction to Machine Learning in Healthcare 1

    1.1 Understanding the Basics 1

    1.2 Methodologies and Applications 4

2. Fundamentals of Deep Learning 7

    2.1 Understanding Neural Networks 7

    2.2 Convolutional Neural Networks (CNN) 17

    2.3 Recurrent Neural Networks (RNNs) 21

3. The Power of Medical Imaging 27

    3.1 Basics of Medical Imaging 27

    3.2 Challenges and Opportunities 38

    3.3 Machine Learning in Medical Imaging 38

4. Diagnosis and Prognosis with ML 45

    4.1 Intro to Predictive Modeling in Healthcare 45

    4.2 Disease Diagnosis Using Machine Learning 47

    4.3 Prognostic Models for Treatment Planning 49

4.4 Cox Proportional Hazards Regression······51

5. Future Trends and Innovations ······56

5.1 Advancements in Healthcare AI Research········56

5.2 Integration of ML into Medical Practice···········59

5.3 Future of Healthcare Technology······················61

Conclusion ······64

1. Introduction to Machine Learning in Healthcare

In healthcare, where data and medical knowledge come together to drive progress, machine learning is a powerful tool. Machine learning is a part of artificial intelligence that teaches computers to learn from patterns in data and get better at tasks over time without needing to be specifically programmed for each task. This is really important in healthcare because there's so much data being generated all the time, and it's coming in really quickly and in many different forms.

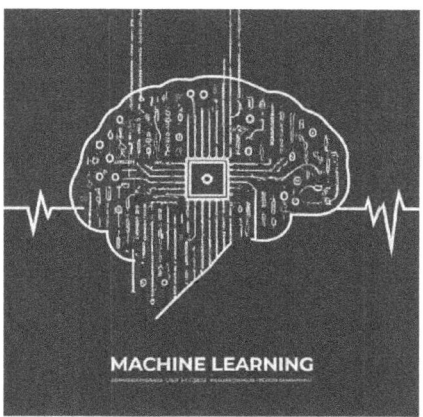

1.1 Understanding the Basics

Machine learning is like a smart tool that helps computers learn from lots of information. There are different types, from simple ones like linear regression to more complex ones like convolutional neural networks (CNNs) and recurrent neural networks (RNNs). These tools can look at big sets of data, find patterns, and give us useful information. In healthcare, they're super helpful because they can help doctors make better decisions, like diagnosing diseases, planning treatments, and figuring out where to put resources.

Let's compare how things usually work in healthcare without machine learning to how they work when we use it. Normally, doctors make decisions based on their own experience, rules they've learned, and small sets of data. But this way has limits because humans can have biases, and we can't process huge amounts of different kinds of data very well.

But when we add machine learning into the mix, it's like giving doctors a superpower. Machine learning can look at tons of medical records, images like X-rays, genetic info, and even data from wearable devices. With all this info, it can help doctors make better predictions about a patient's health, figure out who's at risk for certain diseases, and even suggest personalized treatments.

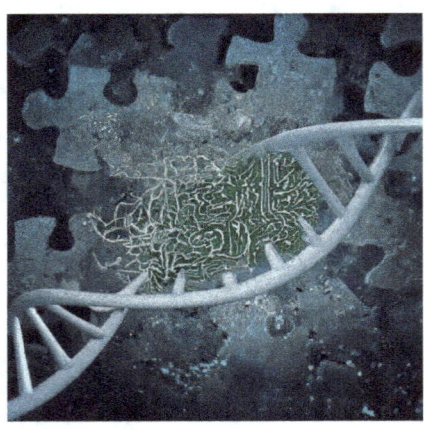

So, why is this so important? Well, recent research shows that using machine learning in healthcare can really improve how we diagnose diseases, predict how they'll progress, find new treatments, and make treatments work better. And with healthcare costs going up and more people needing care, it's more important than ever to use technology like machine learning to make healthcare better for everyone.

Imagine you're at the doctor's office, and they need to figure out if you have a certain disease. Without machine learning, the doctor might rely on what they know from medical school and past patients. But with machine learning, they can use a special computer program that looks at thousands of cases like yours from all around the world. This program can find patterns in the data that the human eye might miss. It can then give the doctor a better idea of what's going on with you and suggest the most effective treatments.

Here's another example: Let's say you have a family history of heart disease, but you want to know your own risk. Without machine learning, it might be hard for your doctor to predict your chances accurately. But with machine learning, they can analyze not only your family history but also other factors like your lifestyle, diet, and exercise habits. By considering all these factors together, the machine learning program can give you a personalized risk assessment, helping you make informed decisions about your health.

Now, let's talk about how machine learning can help with medical imaging, like X-rays or MRIs. Traditionally, doctors would examine these images themselves, which can be time-consuming and prone to human error. But with machine learning, we can teach computers to "see" like doctors. For example, a machine learning algorithm can analyze thousands of X-rays of lungs and learn to spot patterns associated with diseases like pneumonia or lung cancer. This can help radiologists make faster and more accurate diagnoses, leading to better outcomes for patients. These examples illustrate just a few of the ways machine learning is transforming healthcare. By harnessing the power of data and algorithms, we can make medical diagnosis and treatment more accurate, personalized, and efficient than ever before.

1.2 Methodologies and Applications

In the healthcare field, machine learning is like a handy toolbox that helps solve many problems. One tool in this toolbox is deep learning, a special type of machine learning. Let's see how these tools are used and what they do:

1. Medical Imaging Analysis: Imagine looking at a complicated medical image, like an X-ray or MRI. It's not easy for humans to spot all the details, but machines are getting really good at it! Using something called convolutional neural networks (CNNs), machines can analyze these images. They break them down into smaller pieces and learn to recognize patterns. This helps doctors find things like tumors or broken bones more accurately.

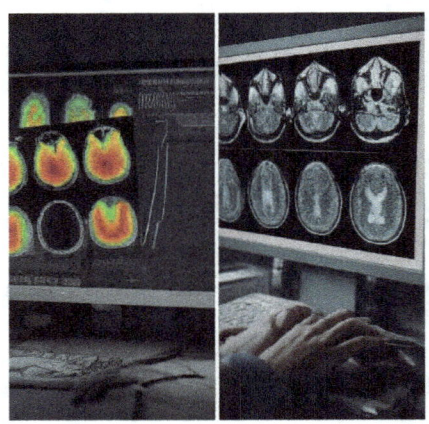

2. Clinical Decision Support Systems: When doctors need to make important decisions about a patient's care, they can use something called a clinical decision support system. This system takes information about the patient, like their medical history and symptoms, and compares it to lots of other data. Machine learning algorithms help these systems make smart suggestions about diagnosis and treatment. It's like having a helpful assistant who knows a lot about medicine.

3. Predictive Modeling: Ever wish you could predict the future? Well, machine learning can't quite do that, but it can make really good guesses! Predictive models use information from a patient's past to guess what might happen in the future. For example, they can guess if someone might get sick again after leaving the hospital. By looking at things like age, health history, and test results, these models can help doctors plan ahead and take action to keep patients healthy.

4. Precision Medicine and Genomic Analysis: Everyone's body is a little different, especially when it comes to our genes. Machine learning helps doctors understand how our genes affect our health.

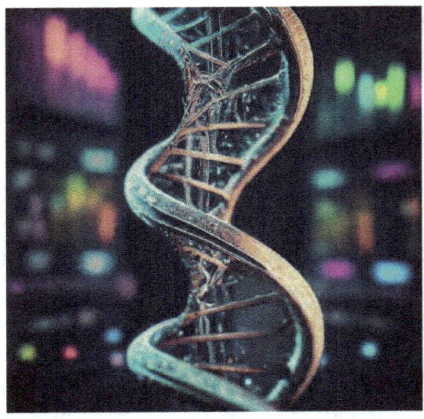

By studying things like DNA sequences and gene patterns, machines can find out if someone is more likely to get certain diseases. This helps doctors choose the best treatments for each person, based on their unique genetic makeup. It's like personalized medicine tailor-made just for you.

These are just a few examples of how machine learning and deep learning are making a big difference in healthcare. With these tools, doctors can diagnose diseases more accurately, plan treatments more effectively, and even prevent illnesses before they happen. It's an exciting time for medicine, and these smart machines are helping us take better care of ourselves and each other.

2. Fundamentals of Deep Learning

In the ever-evolving landscape of healthcare, the emergence of deep learning has heralded a new era of innovation and discovery. Deep learning, a subset of machine learning, revolves around complex neural network architectures inspired by the human brain. In this chapter, we delve into the foundational principles of deep learning and explore its applications within the realm of healthcare.

2.1 Understanding Neural Networks

Imagine a vast network of interconnected nodes, each one working together to solve a complex puzzle. That's a bit like how a neural network works, except instead of nodes, we have artificial neurons, and instead of solving puzzles, they help us solve problems in healthcare.

What are Neural Networks?
Neural networks are computer systems inspired by the human brain. Just like our brains have billions of interconnected neurons, neural networks have layers of artificial neurons that work together to process information. Each neuron takes input, processes it in a special way, and sends an output to other neurons.

Imagine you're in a bustling city, and you're trying to get from one place to another. Along the way, you encounter traffic lights, signs, and other people—all giving you information about where to go and what to do. Your brain processes all this information and helps you navigate your way.

Now, think of a neuron as a tiny messenger in your brain.

It's like a little worker bee buzzing around, carrying information from one place to another. But these "worker bees" aren't just randomly flying around—they're organized in a very specific way to help your brain function efficiently.

Anatomy of a Biological Neuron: A neuron has three main parts: the cell body, dendrites, and axon. The cell body is like the command center—it processes incoming signals and decides what to do with them. Dendrites are like antennas—they receive signals from other neurons and pass them along to the cell body. The axon is like a long wire—it carries signals away from the cell body to other neurons.

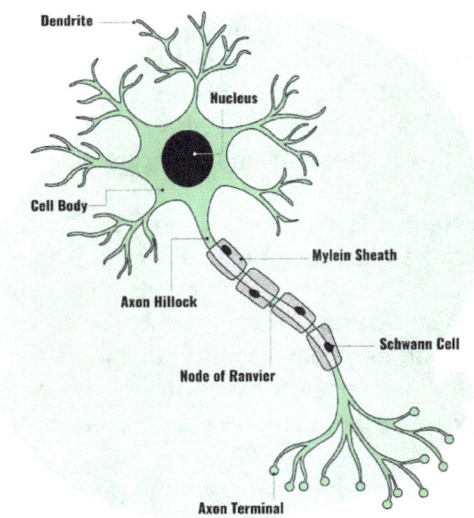

Firing Signals: Neurons communicate with each other through electrical impulses. When a neuron receives enough signals from its dendrites, it gets excited and fires off its own signal. This signal travels down the axon and stimulates other neurons, continuing the chain reaction.

Synapses: Neurons don't physically touch each other. Instead, they're separated by tiny gaps called synapses. When a signal reaches the end of an axon, it releases chemical messengers called neurotransmitters into the synapse. These neurotransmitters then bind to receptors on the dendrites of the next neuron, triggering a new electrical signal.

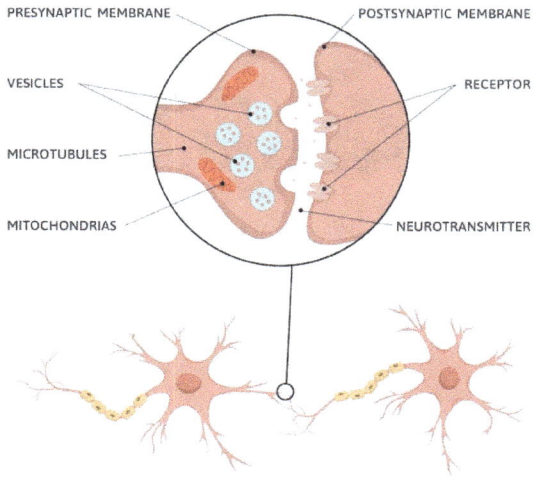

Excitatory vs. Inhibitory Signals: Not all signals are created equal. Some signals make a neuron more likely to fire (excitatory), while others make it less likely to fire (inhibitory). It's like deciding whether to go left or right at a fork in the road—excitatory signals push the neuron in one direction, while inhibitory signals push it in the other.

Learning and Plasticity: Neurons are incredibly adaptable. When you learn something new, like riding a bike or memorizing a song, your neurons form new connections and strengthen existing ones. This process, known as synaptic plasticity, allows your brain to constantly rewire itself and adapt to new experiences.

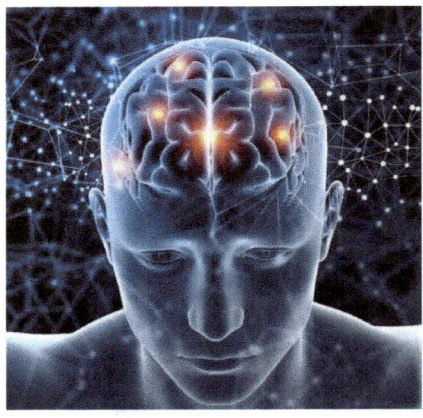

Neurons are the building blocks of the brain, working together to process information, make decisions, and control our actions. By understanding how neurons function, we gain insight into the inner workings of the brain and unlock the mysteries of human cognition and behavior.

Just as biological neurons are the fundamental units of the brain, artificial neurons form the basic building blocks of artificial neural networks (ANNs), the digital counterparts to the human brain. These artificial neurons, also known as perceptrons, play a crucial role in processing information and making decisions in the realm of artificial intelligence.

Anatomy of an Artificial Neuron: While artificial neurons don't have physical structures like cell bodies or dendrites, they consist of mathematical functions designed to mimic the behavior of biological neurons. Each artificial neuron takes in multiple inputs, processes them using mathematical operations, and produces an output signal.

Input Signals: Just like dendrites receiving signals in

biological neurons, artificial neurons receive input signals from other neurons or external sources. These input signals represent various features or attributes of the data being processed, such as pixel values in an image or numerical values in a dataset.

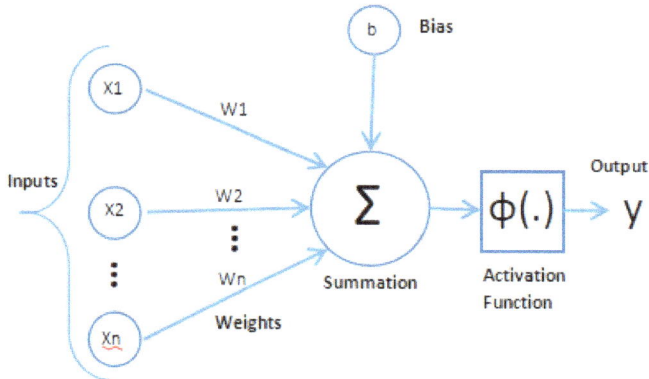

Weights and Bias: One of the key components of artificial neurons is the concept of weights and bias. Each input signal is multiplied by a weight, which determines its importance or relevance to the neuron's output. Additionally, a bias term is added to the weighted sum of inputs to adjust the neuron's activation threshold.

Activation Function: After calculating the weighted sum of inputs and adding the bias, the artificial neuron applies an activation function. This function determines whether the neuron should "fire" or produce an output signal based on the combined input. Common activation functions include the sigmoid function, the hyperbolic tangent (tanh) function, and the rectified linear unit (ReLU) function.

Activation functions are like gatekeepers in artificial neural networks, determining whether a neuron should "fire" and produce an output signal based on its input. These functions play a crucial role in shaping the behavior and learning capabilities of neural networks, enabling them to model complex relationships and make accurate predictions across a wide range of tasks.

Types of Activation Functions:

1. Sigmoid Function: The sigmoid function squashes the input into a range between 0 and 1. It is defined by the formula:

$$\sigma(x) = \frac{1}{1 + e^{-x}}$$

where x is the input to the neuron.

- Pros: It produces a smooth output that is suitable for tasks where the output needs to be interpreted as a probability, such as binary classification.
- Cons: It suffers from the vanishing gradient problem, where the gradients become very small for extreme input values, leading to slower learning.

2. Hyperbolic Tangent Function (tanh): The hyperbolic tangent function is similar to the sigmoid function but squashes the input into a range between -1 and 1. Its formula is:

$$\tanh(x) = \frac{e^x - e^{-x}}{e^x + e^{-x}}$$

- Pros: Like the sigmoid function, it produces a smooth output suitable for tasks requiring bounded outputs. It alleviates the issue of output saturation seen in

the sigmoid function.
- Cons: It still suffers from the vanishing gradient problem.

3. Rectified Linear Unit (ReLU): The ReLU function returns the input as-is if it is positive and sets it to zero otherwise. Its formula is:

$$\mathrm{ReLU}(x) = \max(0, x)$$

- Pros: It is computationally efficient and alleviates the vanishing gradient problem by maintaining a constant gradient for positive inputs.
- Cons: It suffers from the "dying ReLU" problem, where neurons may become inactive (output zero) for negative inputs during training and never recover.

4. Leaky ReLU and Parametric ReLU: To address the issue of "dying" ReLU neurons (i.e., neurons that always output zero for negative inputs), variations of the ReLU function have been proposed. Leaky ReLU allows a small, non-zero gradient for negative inputs, preventing neurons from becoming inactive. Parametric ReLU takes this a step further by allowing the slope of the negative part of the function to be learned during training, providing greater flexibility.

5. Softmax Function: The softmax function is commonly used in the output layer of neural networks for multi-class classification tasks. It converts the raw output scores of the network into probabilities, ensuring that they sum up to 1. This allows the network to output a probability distribution over multiple classes, making it suitable for tasks where the input can belong to one of several categories.

So, we can conclude that Activation functions play a crucial role in shaping the behavior of artificial neurons within neural networks. By choosing the appropriate activation function, we can ensure efficient learning and effective representation of complex data patterns, ultimately leading to improved performance in various machine learning tasks.

Output Signal: Once the activation function is applied, the artificial neuron produces an output signal, which serves as the input to other neurons in the network or as the final output of the network itself. This output signal represents the neuron's response to the input data and plays a crucial role in the overall computation performed by the neural network.

```
output = Sum(weights * inputs) + bias
```

Learning and Adaptability: Similar to biological neurons, artificial neurons are capable of learning and adapting to different tasks through a process known as training. During training, the weights and biases of the neurons are adjusted based on feedback from the desired output, allowing the network to improve its performance over time.

Artificial neurons are the computational counterparts to biological neurons, enabling the creation of powerful artificial neural networks for tasks ranging from image recognition to natural language processing. By understanding the principles of artificial neurons, we gain insight into the inner workings of artificial intelligence and its potential to transform various industries and domains.

Layers of Neurons:
Artificial Neural networks(ANN) are organized into layers, kind of like floors in a building. There are usually three types of layers: input, hidden, and output. The input layer

is where we give the network our data to work with. The hidden layers are where all the magic happens. They process the data and look for patterns. Finally, the output layer gives us the network's answer to our problem.

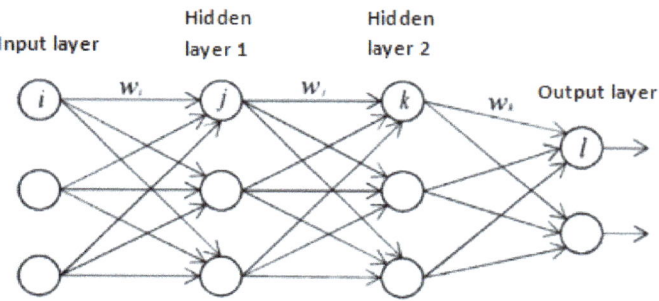

1. **Input Layer:** The input layer is where the network receives data from the outside world. Each neuron in this layer represents a feature or attribute of the input data. For example, in an image recognition task, each neuron might correspond to a pixel in the image. The values of these neurons serve as the initial input to the network and are passed on to the next layer for processing.

2. **Hidden Layers:** Hidden layers are where the magic happens! These layers perform the bulk of the computation in the network, extracting intricate patterns and relationships from the input data. Each neuron in a hidden layer receives inputs from the neurons in the previous layer, applies a transformation using weights and biases, and passes the result to the neurons in the next layer. The number of hidden layers and the number of neurons in each layer are hyperparameters that can be adjusted to optimize the network's performance.

3. **Output Layer:** The output layer is where the network

produces its final predictions or outputs. Each neuron in this layer represents a possible outcome or class label. For example, in a classification task, each neuron might correspond to a different class (e.g., cat, dog, bird). The values produced by the neurons in the output layer are often interpreted as probabilities, indicating the likelihood of each class given the input data.

Propagation Techniques:

1. **Forward Propagation:** Forward propagation is the process by which input data is fed forward through the network, layer by layer, to produce an output prediction. It starts at the input layer, where the input data is received, and progresses through the hidden layers until reaching the output layer. Each neuron in the network applies a transformation to its inputs using weights and biases and passes the result to the neurons in the next layer.

2. **Backward Propagation(Backpropagation):** Backward propagation is the process by which the network learns from its mistakes and adjusts its weights and biases to improve its performance. It starts at the output layer, where the network's predictions are compared to the true labels or targets, and calculates the error or loss. The error is then propagated backward through the network, layer by layer, using techniques such as gradient descent, to update the weights and biases and minimize the error.

Training:
Neural networks are like detectives in training—they learn from their mistakes! We give them lots of examples to practice on, and each time they make a mistake, they learn from it and try to do better next time. This process is called training, and it's how neural networks get smarter over

time.

Imagine our detective not just solving one mystery but multiple mysteries at the same time! Deep learning is all about using multiple layers of neurons to solve really complex problems. Each layer learns to recognize different features in the data, like edges in an image or words in a sentence.

2.2 Convolutional Neural Networks (CNN)

Imagine you're looking at a painting. You don't see the whole picture at once; instead, you focus on small parts, like the brush strokes or the colors. Convolutional Neural Networks (CNNs) work a bit like this. They're specialized for processing images, breaking them down into smaller, manageable pieces to understand them better.

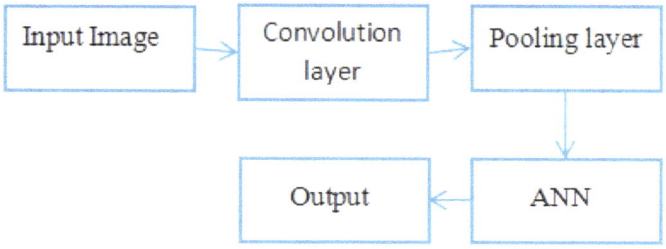

At the heart of a CNN are layers of neurons, just like in our brain. But these neurons aren't looking at the whole picture; instead, they focus on tiny sections called "*filters*" or "*kernels*." Each filter looks for specific features in the image, like edges, corners, or textures.

When we talk about "*convolution*," we're basically sliding these filters across the image, looking at one small piece at a time. As we move the filter, we multiply its values with the pixel values in the image and add them up. This creates a new, filtered version of the image that highlights certain features.

1. Input Image:
The journey of a Convolutional Neural Network (CNN) begins with the input image. This could be any digital image, such as a medical scan, a photograph, or a digital artwork. The image is typically represented as a grid of pixels, with each pixel containing numerical values representing its color intensity or grayscale value.

2. Convolutional Layer:
The convolutional layer is the core building block of a CNN. It consists of a set of filters, also known as kernels, that slide over the input image. Each filter detects specific features, such as edges, textures, or patterns, within the image. As the filters move across the image, they perform a mathematical operation called convolution, which involves multiplying the filter values with the pixel values in the image and summing them up. This process creates a feature map that highlights where certain features are present in the image.

3. Activation Function:
After the convolutional operation, each neuron in the feature map passes its output through an activation function. One commonly used activation function is the Rectified Linear Unit (ReLU), which introduces non-linearity into the network by converting negative values to zero. This helps the network learn complex relationships and make more accurate predictions.

4. Pooling Layer:
The pooling layer is responsible for reducing the spatial dimensions of the feature maps while retaining the most important information. This is typically achieved through operations like max pooling or average pooling, where the

network selects the maximum or average value from a small region of the feature map. Pooling helps make the representations more manageable and invariant to small variations in the input, making the network more robust and efficient.

5. Fully Connected (FC) Layer (Artificial Neural Network - ANN):

The fully connected layer, also known as the dense layer, is a traditional neural network layer where each neuron is connected to every neuron in the previous layer. In the context of a CNN, the flattened output from the previous layers (after pooling) is fed into the fully connected layer. This layer learns complex combinations of features from the input image and is responsible for making final predictions or classifications. It often employs activation functions like ReLU or softmax for classification tasks.

6. Output Layer:

Finally, the output layer of the CNN produces the network's predictions or classifications based on the features learned from the input image. The number of neurons in the output layer depends on the specific task the CNN is designed for. For example, in a binary classification task (e.g., tumor detection), there may be one neuron representing each class (tumor or non-tumor), while in a multi-class classification task (e.g., identifying different types of tumors), there may be multiple neurons, each corresponding to a different class. The output layer typically uses an appropriate activation function, such as sigmoid for binary classification or softmax for multi-class classification, to produce probability scores for each class.

The layers of a CNN work together to extract meaningful features from the input image, reduce its spatial dimensions, learn complex patterns, and make predictions or classifications based on the learned features. This hierarchical architecture enables CNNs to effectively analyze images and perform a wide range of tasks in various domains.

Transfer Learning:

Transfer learning in Convolutional Neural Networks (CNNs) is a technique where a pre-trained model, which has been trained on a large dataset for a specific task, is adapted or fine-tuned to perform a different but related task. Instead of starting the training process from scratch, transfer learning leverages the knowledge gained from the pre-trained model and applies it to the new task, typically requiring less training data and less computational resources.

Here's how transfer learning works in CNNs:

Pre-Trained Model: A CNN model is trained on a large dataset for a particular task, such as image classification. This pre-trained model has learned to extract general features from images, like shapes, textures, and patterns.

Task Adaptation: When faced with a new task, such as a different classification problem or a specific domain like medical imaging, transfer learning allows us to adapt the pre-trained model to this new task. Instead of training the model from scratch, we modify the existing model's architecture or fine-tune its parameters to better suit the new task.

Feature Extraction: In transfer learning, the learned representations (features) from the pre-trained model's convolutional layers are often preserved and used as inputs to a new set of fully connected layers, which are then trained specifically for the new task. This approach allows the model to focus on learning task-specific features while leveraging the general features learned from the pre-trained model.

Fine-Tuning: Additionally, we may fine-tune the entire pre-trained model or certain layers of it on the new task. Fine-tuning involves adjusting the parameters of the pre-trained model using a smaller dataset specific to the new task. This step helps the model adapt to the nuances of the new data and further improves its performance on the target task.

Transfer learning offers several advantages, including:
Faster Training: Since the model starts with pre-learned features, it requires less time and computational resources to train on the new task.
Better Generalization: Transfer learning helps improve the generalization ability of the model, especially when the new task has a limited amount of training data.
Domain Adaptation: It allows models trained on one domain (e.g., general images) to be adapted to another domain (e.g., medical images) with minimal effort.

Overall, transfer learning is a powerful technique in CNNs that enables the reuse of knowledge learned from one task to benefit another, facilitating faster model development and improving performance, especially in scenarios with limited data availability.

2.3 Recurrent Neural Networks (RNNs)

Imagine you're reading a book. As you go from one page to the next, the story unfolds gradually. Now, what if I told you that your brain works a bit like that when it's processing information? That's where recurrent neural networks (RNNs) come in.

What Are Recurrent Neural Networks?
RNNs are a special type of neural network designed to handle sequences of data. It s like a brainy language expert in the world of computers. It's a special kind of deep

learning tool that not only understands the words you feed it but also remembers what came before and predicts what might come next.

Why RNN?
RNNs are superheroes when it comes to handling sequences of data. Imagine you're typing a text message: "I like eating Pizzas. My favorite is margarita____". For us humans, it's easy to guess the missing word ("pizza"), right? But for a machine, it's a bit trickier. It needs to understand the context and remember the words that came before to make a good guess. That's where RNNs come to the rescue!

How Do RNNs Work?
Think of an RNN as a series of interconnected cells, each passing information to the next one in the sequence. When it receives an input (let's say a word in a sentence), it processes it and produces an output (maybe predicting the next word). But here's the cool part: the RNN also keeps a memory of what it's seen before. So, when it sees the next word, it can use that memory to make a better prediction.

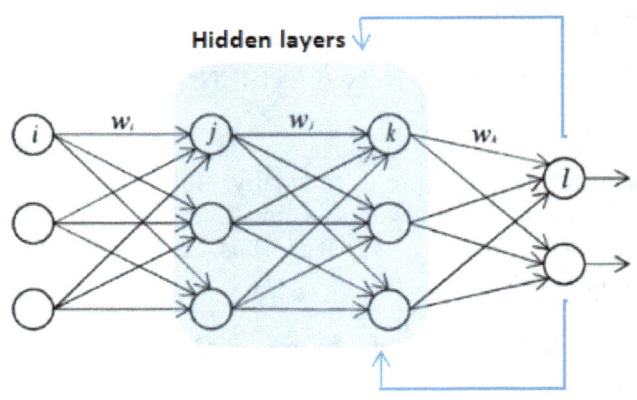

Now, let's get a bit technical. In an RNN, each cell has two main components: the input, which is the data it receives at each time step, and the hidden state, which is like its memory of past inputs. Here's how it works:

- Hidden State: This is the memory of the cell at a particular time step. It's calculated based on the input at that time step and the previous hidden state, using a set of weights and activation functions.
- Output: This is the prediction or output of the cell at each time step, based on the current hidden state.
- Activation Function (σ): This is a mathematical function that introduces non-linearity into the RNN, allowing it to capture complex patterns in the data.

Types of Recurrent Neural Networks (RNNs)
Just like there are different genres of books, there are also different types of RNN architectures, each suited for specific tasks. Here are a few common ones:

Vanilla RNNs: These are the simplest form of RNNs, where each cell passes its hidden state to the next one in the sequence. However, they suffer from the vanishing gradient problem, which makes it hard for them to capture long-term dependencies in the data.

Long Short-Term Memory (LSTM) Networks: LSTMs are an improved version of RNNs designed to address the vanishing gradient problem. They have additional mechanisms called gates that control the flow of information through the network, allowing them to remember important information over longer sequences.

Gated Recurrent Units (GRUs): GRUs are similar to LSTMs but have a simpler architecture with fewer parameters. They're often used when you need a balance between performance and simplicity.

Let's put all this into context with a real-world example. Imagine you're a doctor trying to predict whether a patient with a history of heart disease will have a heart attack in the next year. You have access to their medical records, including things like blood pressure, cholesterol levels, and previous heart-related incidents.

Now, you could use an RNN to analyze this data over time. Each time step represents a month or a year, and at each step, the RNN takes in the patient's current health metrics and predicts the likelihood of a future heart attack. But it doesn't just look at the current data—it also remembers past patterns and trends, like how the patient's blood pressure has been trending over time.

By analyzing this sequential data, the RNN can help you make more informed decisions about the patient's care, whether it's adjusting their medication, recommending lifestyle changes, or scheduling regular check-ups.

2.4 Applications of Deep Learning in Healthcare

Imagine it like having a super-powered assistant who can look at a bunch of medical data and find hidden patterns or make smart predictions. Let's explore how deep learning comes to the rescue in healthcare:

1. Medical Imaging Analysis: When doctors need to look inside our bodies to diagnose illnesses or injuries, they use medical imaging techniques like X-rays, MRIs, or CT scans. Deep learning algorithms, especially Convolutional Neural Networks (CNNs), are like expert detectives trained to analyze these images. They can spot things like tumors, fractures, or abnormalities that might be hard for humans to see. It's like having an extra pair of eyes that never get tired.

2. Disease Diagnosis and Prognosis: Deep learning isn't just about looking at pictures; it's also great at diagnosing diseases and predicting how they might progress. Imagine you're feeling unwell, and you visit your doctor. They run some tests and gather information about your symptoms. Deep learning algorithms can analyze all this data and help your doctor figure out what's wrong with you. They might even predict how your illness could develop in the future, helping your doctor plan the best treatment for you.

3. Drug Discovery and Development: Creating new medicines is a bit like searching for a needle in a haystack. It takes a lot of time and effort to find a compound that can treat a specific disease. Deep learning algorithms can speed up this process by analyzing huge datasets of chemical compounds and predicting which ones might be effective against certain diseases. It's like having a super-smart assistant in the lab, helping researchers find new cures faster.

4. Personalized Medicine: We're all unique, and what works for one person might not work for another. Deep learning helps doctors tailor treatments to each individual's needs based on their genetic makeup, medical history, and lifestyle. By analyzing vast amounts of data, deep learning algorithms can predict how different patients might respond to various treatments. This personalized approach to medicine ensures that each patient gets the care that's right for them.

5. Healthcare Management and Operations: By analyzing data from electronic health records, deep learning algorithms can optimize hospital workflows, predict patient admissions, and even detect fraud or errors in billing. It's like having a smart organizer behind the scenes, making sure everything runs like clockwork.

6. Remote Patient Monitoring and Telemedicine: With remote patient monitoring and telemedicine, patients can receive care from the comfort of their homes. Deep learning algorithms play a crucial role here by analyzing data from wearable devices, such as smartwatches or fitness trackers, to monitor patients' health in real-time. They can detect anomalies or changes in vital signs, alerting healthcare providers to intervene if necessary. This technology allows patients to stay connected with their doctors and receive timely care, even from miles away.

7. Public Health Surveillance: When it comes to keeping communities healthy, early detection is key. Deep learning algorithms can sift through vast amounts of data, including social media posts, internet searches, and health records, to identify potential disease outbreaks or public health trends. By analyzing patterns and detecting signals, these algorithms help public health officials take proactive measures to prevent the spread of diseases and protect the population. It's like having a virtual watchdog that keeps an eye on the health of entire communities, helping to stop outbreaks before they become widespread.

8. Rehabilitation and Assistive Technologies: For individuals with disabilities or injuries, rehabilitation and assistive technologies can make a world of difference in restoring mobility and independence. Algorithms can analyze motion data from sensors or cameras to track patients' movements and provide personalized feedback or assistance. Whether it's helping stroke patients regain motor function or assisting individuals with prosthetic limbs, deep learning technologies empower individuals to live fuller, more active lives.

3. The Power of Medical Imaging

Medical imaging plays a pivotal role in modern healthcare by providing doctors with essential insights into the human body's internal structures and functions. From X-rays to MRIs, these imaging techniques enable healthcare professionals to diagnose diseases, monitor treatment progress, and plan interventions. However, the interpretation of medical images can be complex and subjective, relying heavily on the expertise of trained professionals.

In recent years, the integration of machine learning into medical imaging has revolutionized the field, offering new avenues for improving accuracy, efficiency, and diagnostic outcomes. Machine learning algorithms, powered by vast amounts of medical data, can learn patterns and features within images that may not be immediately apparent to the human eye. By automating tasks such as image segmentation, feature extraction, and disease classification, machine learning holds the promise of streamlining the diagnostic process and enhancing patient care.

3.1 Basics of Medical Imaging

Medical imaging encompasses various techniques that enable healthcare professionals to visualize the internal structures of the human body. Each modality utilizes different principles to generate images, providing unique insights into different aspects of anatomy and pathology.

Overview of Medical Imaging Modalities:

a) X-ray:
X-ray imaging, also known as radiography, is a widely

used medical imaging technique that employs electromagnetic radiation to produce images of the internal structures of the body. It relies on the principle that different tissues absorb X-rays to varying degrees, resulting in variations in the intensity of the X-ray beam that reaches the detector.

Principles of X-ray Imaging:

X-ray Generation: X-rays are produced when high-energy electrons collide with a metal target, typically tungsten, within an X-ray tube. This process generates a beam of X-rays that passes through the body.

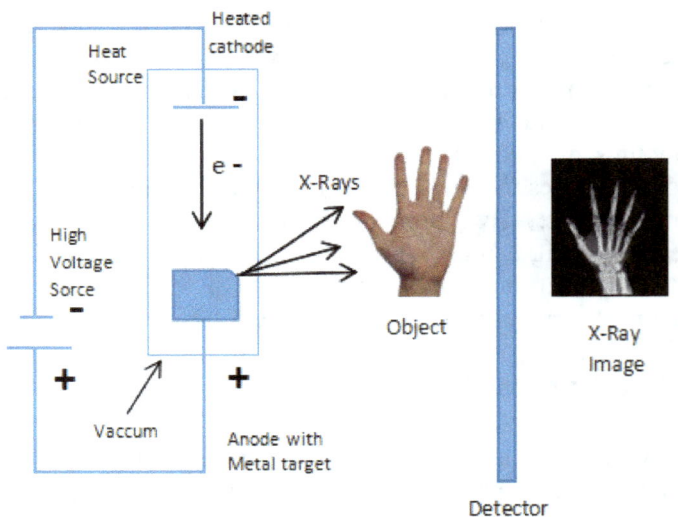

Absorption and Transmission: As the X-ray beam traverses the body, it interacts with the tissues it encounters. Dense tissues, such as bones, absorb more X-rays and appear white on the resulting image (radiopaque), while less dense tissues, such as muscles and organs, allow more X-rays to pass through and appear darker (radiolucent).

Image Formation: The X-ray beam that passes through the body is captured by a detector, such as a photographic film or a digital sensor. The resulting image, known as a radiograph, depicts the internal structures of the body in two dimensions.

Applications of X-ray Imaging:
Bone Fractures and Trauma: X-rays are commonly used to diagnose fractures, dislocations, and other orthopedic injuries. They provide detailed views of bones and can help determine the extent and location of fractures.

Chest Imaging: X-rays are valuable for evaluating the lungs and chest cavity, enabling the detection of conditions such as pneumonia, lung cancer, and pleural effusion.

Dental Imaging: In dentistry, X-rays are used to assess the health of teeth, detect cavities, and evaluate the supporting structures of the jaw.

Diagnostic Screening: X-ray imaging may be employed as a screening tool for certain medical conditions, such as tuberculosis or scoliosis.

Advantages and Limitations:
Advantages: X-ray imaging is widely available, relatively inexpensive, and provides quick results. It is particularly effective for visualizing dense structures, such as bones.

Limitations: X-rays expose patients to ionizing radiation, which carries a small risk of harm, particularly with repeated exposures. Additionally, X-ray images may lack detail in soft tissues and may not always provide sufficient information for certain diagnostic purposes.

Despite its limitations, X-ray imaging remains an indispensable tool in medical diagnostics, offering valuable insights into a wide range of conditions and facilitating timely and accurate diagnoses. Advances in technology, such as digital radiography and computed radiography, continue to enhance the quality and efficiency of X-ray imaging procedures.

b) Computed Tomography (CT):

Computed Tomography (CT), also known as computed axial tomography (CAT) scanning, is a sophisticated medical imaging technique that provides detailed cross-sectional images of the body. CT scans utilize a combination of X-rays and computer processing to produce high-resolution images that offer valuable diagnostic information.

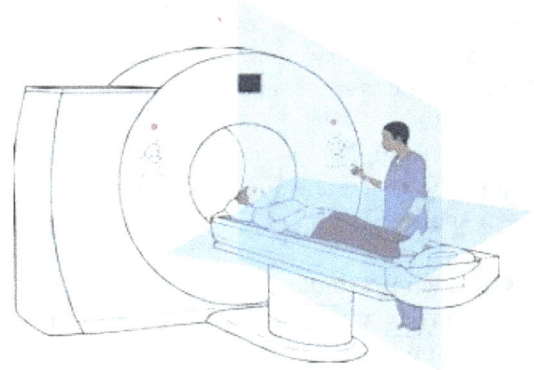

Principle of Operation:

During a CT scan, the patient lies on a motorized table that moves through a doughnut-shaped machine called a CT scanner. This scanner consists of an X-ray tube that rotates around the patient, emitting narrow beams of X-rays from multiple angles.

Detectors opposite the X-ray source measure the amount of radiation that passes through the body at each angle. These measurements are then processed by a computer to reconstruct cross-sectional images, or "slices," of the body. By capturing images from different perspectives, CT scans create detailed three-dimensional representations of internal structures, allowing healthcare professionals to visualize organs, tissues, and abnormalities with exceptional clarity.

Applications:
- CT scans are versatile diagnostic tools used in various medical specialties, including radiology, emergency medicine, oncology, and surgery.
- They are particularly valuable for evaluating conditions such as traumatic injuries (e.g., fractures, internal bleeding), acute abdominal pain (e.g., appendicitis, bowel obstruction), and diseases of the chest (e.g., lung cancer, pulmonary embolism).
- CT angiography (CTA) is a specialized application of CT scanning used to visualize blood vessels and assess vascular conditions such as arterial blockages, aneurysms, and arteriovenous malformations.

Advantages:
- CT scans provide detailed anatomical information with excellent spatial resolution, making them effective for detecting small lesions and abnormalities.
- They are faster than traditional X-ray imaging and can capture images of multiple body regions in a single examination.

Considerations:
- CT scans involve exposure to ionizing radiation, which carries a small risk of radiation-induced cancer, particularly with repeated or high-dose examinations.
- Special precautions may be necessary for certain patient populations, such as pregnant women and individuals with kidney impairment, to minimize potential risks associated with contrast agents used in

some CT procedures.

In summary, Computed Tomography (CT) is a powerful imaging modality that provides detailed anatomical information across multiple body regions, aiding in the diagnosis and management of a wide range of medical conditions. Its ability to produce high-resolution images quickly and non-invasively makes it an indispensable tool in modern healthcare.

c) Magnetic Resonance Imaging (MRI):
Magnetic Resonance Imaging (MRI) is a sophisticated medical imaging technique that utilizes a powerful magnetic field, radio waves, and computer technology to produce detailed images of the internal structures of the body. Unlike X-rays and CT scans, which use ionizing radiation, MRI relies on the natural magnetic properties of atoms within the body, primarily hydrogen atoms found in water and fat molecules.

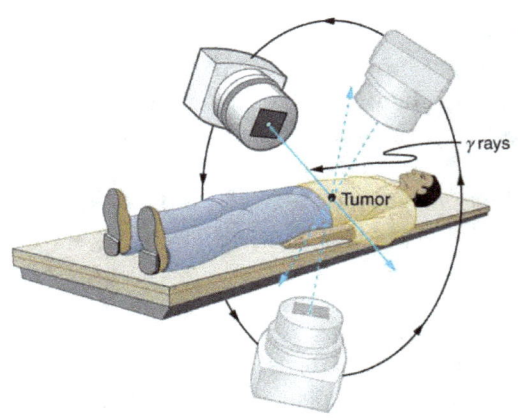

Principles of MRI:
- When a patient is placed inside the MRI scanner, the magnetic field aligns the hydrogen atoms within their

body along its direction.
- Radiofrequency pulses are then applied to the body, causing the hydrogen atoms to absorb energy and temporarily deviate from their aligned position.
- As the atoms return to their original alignment, they emit signals that are detected by specialized antennas in the MRI machine.
- These signals are processed by computer algorithms to construct detailed two-dimensional or three-dimensional images of the body's internal structures.

Advantages of MRI:
- **Excellent Soft Tissue Contrast:** MRI provides exceptional contrast resolution, making it particularly suitable for imaging soft tissues such as the brain, spinal cord, muscles, and organs.
- **Non-ionizing Radiation:** Unlike X-rays and CT scans, MRI does not use ionizing radiation, reducing the risk of radiation exposure and making it safer for patients, especially children and pregnant women.
- **Multi-planar Imaging:** MRI can acquire images in multiple planes (axial, sagittal, and coronal), allowing for comprehensive evaluation of anatomical structures from different perspectives.
- **Functional Imaging:** Advanced MRI techniques, such as functional MRI (fMRI) and diffusion tensor imaging (DTI), can assess brain function, connectivity, and microstructural integrity, enabling researchers and clinicians to study neurological disorders and brain function in vivo.

Clinical Applications of MRI:
Neuroimaging: MRI is widely used to diagnose and monitor neurological conditions such as stroke, brain tumors, multiple sclerosis, and traumatic brain injury.
Orthopedics: MRI is valuable for evaluating musculoskeletal disorders, including ligament and tendon

injuries, joint abnormalities, and degenerative diseases such as osteoarthritis.

Oncology: MRI plays a crucial role in cancer diagnosis, staging, and treatment planning, offering detailed information about tumor size, location, and involvement of adjacent structures.

Cardiology: MRI can assess cardiac structure and function, detect myocardial infarction, and evaluate congenital heart defects, providing valuable insights into cardiovascular health.

d) Ultrasound:

Ultrasound imaging, also known as sonography, is a non-invasive medical imaging technique that utilizes high-frequency sound waves to produce real-time images of internal structures within the body. Unlike other imaging modalities such as X-ray, CT, or MRI, ultrasound does not involve ionizing radiation, making it a safe and widely used imaging tool, particularly during pregnancy and in pediatric patients.

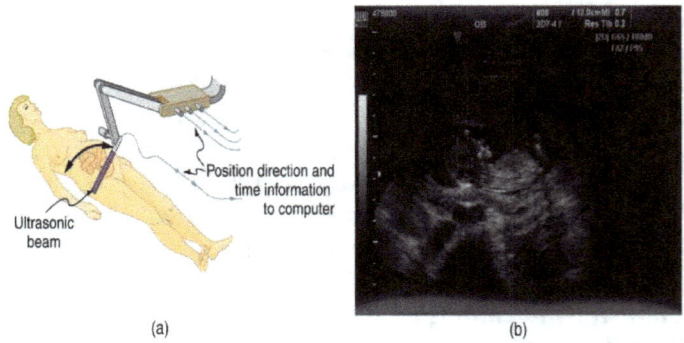

(a) An ultrasonic image is produced by sweeping the ultrasonic beam across the area of interest, in this case the woman's abdomen.. (b) Ultrasound image of 12-week-old fetus. (credit: Margaret W. Carruthers, Flickr)

Principle of Ultrasound:
Ultrasound works on the principle of sound wave reflection. A transducer, which emits and receives ultrasound waves, is placed on the skin surface and directed towards the area of interest. When the ultrasound waves encounter boundaries between different tissues, such as between fluid and soft tissue or soft tissue and bone, some of the waves are reflected back to the transducer. The transducer detects these reflected waves and converts them into electrical signals, which are then processed to create real-time images on a monitor.

Applications of Ultrasound:
Ultrasound imaging has a wide range of applications across various medical specialties:

- **Obstetrics and Gynecology:** Ultrasound is commonly used during pregnancy to monitor fetal development, assess fetal well-being, and detect any abnormalities or complications, such as fetal growth restriction or placental abnormalities. It is also used for gynecological evaluations, such as detecting ovarian cysts or evaluating the uterus and ovaries for abnormalities.
- **Cardiology:** In cardiology, ultrasound, specifically echocardiography, is used to visualize the structure and function of the heart. It allows cardiologists to assess heart valves, chambers, and blood flow patterns, diagnose heart conditions such as heart failure or heart valve disorders, and monitor cardiac function during procedures like stress testing or cardiac catheterization.
- **Abdominal Imaging:** Ultrasound is valuable for evaluating abdominal organs such as the liver, gallbladder, pancreas, kidneys, and spleen. It can detect abnormalities such as gallstones, liver masses,

kidney stones, or abdominal fluid collections. Ultrasound-guided procedures, such as biopsies or fluid drainage, are also performed in the abdomen.
- **Musculoskeletal Imaging:** Ultrasound is increasingly used in musculoskeletal imaging to assess soft tissue and joint abnormalities, such as tendon tears, ligament injuries, muscle strains, or joint effusions. It provides real-time visualization of dynamic movements, making it useful for guiding injections or aspirations in joints or soft tissues.
- **Vascular Imaging:** Doppler ultrasound, a specialized technique that evaluates blood flow, is used to assess vascular conditions such as deep vein thrombosis (DVT), peripheral artery disease (PAD), carotid artery stenosis, or varicose veins. It helps in diagnosing blood clots, evaluating blood flow abnormalities, and assessing the integrity of blood vessels.

Advantages of Ultrasound:
- Non-invasive and painless
- Does not involve ionizing radiation, making it safe for use during pregnancy and in pediatric patients
- Real-time imaging allows for dynamic assessment of structures and movements
- Portable and versatile, suitable for use at the bedside or in various clinical settings
- Cost-effective compared to other imaging modalities like MRI or CT.

In summary, ultrasound imaging is a valuable diagnostic tool with diverse applications across multiple medical specialties. Its non-invasive nature, real-time imaging capabilities, and safety profile make it an indispensable tool for healthcare providers in diagnosing and managing a wide range of conditions.

Image Acquisition and Reconstruction:
Medical imaging encompasses various modalities, each with its unique principles of image acquisition and reconstruction. Traditional methods rely on predefined algorithms and manual intervention for image processing and analysis. However, the advent of deep learning and machine learning has introduced novel approaches to enhance image interpretation by enabling machines to understand patterns across different modalities.

Image Acquisition Principles: Medical images are generated through a range of modalities, including X-rays, CT scans, MRIs, and ultrasounds. Each modality employs distinct physical principles to capture images of internal structures. For example, X-rays use ionizing radiation to produce two-dimensional images, while MRIs utilize magnetic fields and radio waves to generate detailed three-dimensional representations. Understanding these principles is crucial for optimizing image acquisition parameters and ensuring diagnostic accuracy.

Image Reconstruction Techniques: After acquisition, medical images undergo reconstruction to generate visual representations of anatomical structures. Traditional methods involve computationally intensive algorithms to process raw data and create interpretable images. However, deep learning techniques, such as convolutional neural networks (CNNs), have emerged as powerful tools for image reconstruction. CNNs can learn intricate patterns and features within medical images, enabling more accurate and efficient reconstruction processes.

Machine Learning for Image Analysis: Once images are reconstructed, machine learning algorithms can be employed for comprehensive image analysis. These algorithms learn from vast datasets of labeled images to recognize patterns indicative of various pathologies or anatomical structures. By leveraging advanced machine learning techniques, such as deep learning architectures, machines can extract relevant features, classify abnormalities, and assist radiologists in clinical decision-making.

3.2 Challenges and Opportunities

While the integration of deep learning and machine learning into medical imaging holds immense promise, several challenges must be addressed. These include data heterogeneity, model interpretability, and regulatory considerations. Moreover, the ethical implications of automated image analysis, such as patient privacy and algorithmic bias, necessitate careful scrutiny. Nevertheless, the synergy between medical imaging and machine learning offers unprecedented opportunities to enhance diagnostic accuracy, improve patient outcomes, and advance the field of healthcare.

3.3 Machine Learning in Medical Imaging

Intersection of Machine Learning and Medical Imaging:

Medical imaging generates vast amounts of data that can be overwhelming for human interpretation alone. This is where machine learning steps in, offering powerful tools to extract meaningful insights from these images. At its core, machine learning involves teaching computers to learn from data and make predictions or decisions without explicit programming.

In the realm of medical imaging, machine learning algorithms can analyze images with remarkable speed and precision, assisting radiologists and clinicians in their diagnostic tasks. By training on labeled datasets containing examples of normal and abnormal images, these algorithms can learn to recognize subtle patterns indicative of various pathologies.

The intersection of machine learning and medical imaging has led to several breakthroughs. For instance, deep learning techniques such as convolutional neural networks (CNNs) have demonstrated remarkable performance in tasks like image segmentation, where they delineate structures of interest within an image, such as tumors or organs. Similarly, these algorithms excel in feature extraction, automatically identifying relevant characteristics within images that are crucial for diagnosis. Moreover, machine learning enables the development of predictive models that forecast patient outcomes or treatment responses based on imaging data. By analyzing patterns across thousands of images and patient records, these models can identify early indicators of disease progression or treatment efficacy, empowering clinicians to intervene proactively.

Enhancing Image Processing with Machine Learning Algorithms:

Medical imaging generates vast amounts of data, and extracting meaningful information from these images is crucial for accurate diagnosis and treatment planning. Traditional image processing techniques often rely on predefined rules and heuristics, which may not capture the full complexity of medical images or adapt well to variations in patient anatomy and pathology.

Machine learning algorithms offer a more flexible and data-driven approach to image processing in medical imaging. By training on large datasets of annotated medical images, these algorithms can learn complex patterns and relationships, enabling them to perform tasks such as image enhancement, denoising, and artifact correction.

One area where machine learning has shown particular promise is in image reconstruction. In modalities like MRI and CT, raw image data is often noisy and incomplete, requiring sophisticated algorithms to reconstruct high-quality images. Machine learning techniques, such as deep learning-based image reconstruction, have been developed to improve the quality and fidelity of reconstructed images, leading to sharper, clearer images with reduced artifacts.

Additionally, machine learning algorithms can be used for image segmentation, which involves partitioning an image into regions of interest. Segmentation is crucial for tasks such as tumor delineation, organ delineation, and anatomical structure localization. Traditional segmentation algorithms often require manual intervention and tuning of parameters, making them time-consuming and prone to error. Machine learning-based segmentation approaches, on the other hand, can learn to automatically delineate structures from medical images, reducing the need for manual intervention and improving accuracy.

Moreover, machine learning algorithms can aid in feature extraction from medical images, identifying relevant anatomical and pathological features that may be indicative of disease. These features can then be used as input to diagnostic models for disease classification, prognosis, and treatment planning.

Applications of Machine Learning in Medical Image Analysis:

1. Image Segmentation: Image segmentation involves partitioning an image into multiple segments or regions to facilitate further analysis. Algorithms can automatically identify and delineate structures of interest within medical images, such as organs, tumors, or anatomical landmarks.

By segmenting images accurately, clinicians can extract quantitative measurements, track disease progression, and localize abnormalities with precision.

2. Feature Extraction: Feature extraction entails identifying distinctive patterns or characteristics within medical images that are indicative of specific pathologies or conditions. Machine learning algorithms can automatically extract relevant features from images, such as texture, shape, or intensity, which serve as discriminative markers for disease diagnosis or prognosis. These extracted features can aid in differentiating between healthy and diseased tissues, guiding clinicians in making informed decisions regarding patient care.

3. Disease Classification: Disease classification involves categorizing medical images into different classes or diagnostic categories based on their visual characteristics. Machine learning algorithms can learn discriminative patterns from labeled image datasets, enabling automated classification of images into relevant disease categories or subtypes. By accurately classifying medical images, machine learning facilitates early detection of diseases, risk stratification, and personalized treatment planning, ultimately improving patient outcomes.

4. Image Registration: Image registration involves aligning images obtained from different modalities (e.g., MRI, CT) or acquired at different time points to enable meaningful comparison and analysis. Machine learning techniques can facilitate automated image registration by learning complex spatial transformations that align corresponding anatomical structures across images, enhancing the accuracy and efficiency of multi-modal image fusion and longitudinal studies.

5. Image Synthesis: Image synthesis refers to the generation of synthetic medical images with desired characteristics, such as variations in pathology, imaging parameters, or patient demographics. Machine learning models, particularly generative adversarial networks (GANs), can learn underlying distributions of medical image data and generate realistic synthetic images that augment limited datasets, facilitate data augmentation for training deep learning models, and simulate rare or challenging clinical scenarios for educational or research purposes.

6. Anomaly Detection: Anomaly detection involves identifying outliers or abnormalities within medical image datasets that deviate from normal anatomical structures or expected imaging appearances. Machine learning algorithms, including unsupervised learning approaches such as autoencoders or one-class classification techniques, can learn representations of normal image patterns and detect deviations indicative of pathology or artifacts. Anomaly detection aids in quality control of medical imaging datasets, flagging suspicious findings for further review, and improving diagnostic confidence.

Advantages of Machine Learning in Medical Imaging

Machine learning offers a multitude of advantages in the field of medical imaging, revolutionizing traditional approaches to image analysis and interpretation. Here are some key advantages:

- **Automation and Efficiency:** Machine learning algorithms automate labor-intensive tasks in medical image analysis, such as segmentation, feature extraction, and disease classification. This automation reduces the burden on radiologists and clinicians, allowing them to focus their expertise on more complex cases and clinical decision-making. Additionally, machine learning accelerates the processing of large volumes of imaging data, leading to faster diagnosis and treatment planning.

- **Improved Accuracy and Consistency:** Machine learning algorithms leverage patterns and features within medical images that may be imperceptible to the human eye, leading to enhanced accuracy and consistency in image interpretation. By learning from vast datasets, machine learning models can identify subtle abnormalities or variations in imaging findings, reducing diagnostic errors and improving patient outcomes. Moreover, machine learning facilitates standardized interpretation of images across different healthcare settings, ensuring consistent quality of care.

- **Personalized Medicine:** Machine learning enables the development of personalized diagnostic and treatment strategies tailored to individual patient characteristics and preferences. By analyzing diverse patient data, including imaging studies, genetic information, and clinical histories, machine learning algorithms can identify patient-specific risk factors, predict treatment responses, and recommend personalized interventions. This personalized approach to medicine maximizes therapeutic efficacy, minimizes adverse effects, and optimizes healthcare resource utilization.

- **Enhanced Image Quality and Reconstruction:** Machine learning techniques can improve the quality and resolution of medical images, enhancing visualization of anatomical structures and pathological findings. Deep learning-based image reconstruction algorithms can generate high-fidelity images from sparse or noisy data, reducing artifacts and improving image clarity. Additionally, machine learning models can denoise images, correct distortions, and enhance contrast, leading to superior diagnostic accuracy and confidence.

- **Discovery of Novel Biomarkers and Imaging Signatures:** Machine learning facilitates the discovery of novel biomarkers and imaging signatures associated with disease pathology, progression, and treatment response. By analyzing multi-dimensional imaging data, machine learning algorithms can identify subtle imaging biomarkers indicative of early disease stages, prognostic outcomes, or treatment efficacy. These imaging signatures provide valuable insights into disease mechanisms, guide therapeutic development, and enable precision medicine approaches.

- **Scalability and Generalization:** Machine learning models exhibit scalability and generalization capabilities, allowing them to analyze diverse imaging modalities, anatomical regions, and disease conditions. Once trained on representative datasets, machine learning algorithms can generalize their learned knowledge to unseen data, enabling seamless integration into clinical workflows and adaptation to new healthcare challenges. This scalability ensures the widespread adoption and applicability of machine learning in medical imaging across different healthcare settings and patient populations.

Overall, machine learning holds immense promise for advancing the field of medical imaging, empowering healthcare professionals with powerful tools for diagnosis, prognosis, and treatment optimization. By harnessing the advantages of machine learning, we can unlock the full potential of medical imaging technologies to improve patient care and reshape the future of healthcare.

4. Diagnosis and Prognosis with ML

4.1 Intro to Predictive Modeling in Healthcare

Predictive modeling in healthcare constitutes a multifaceted approach aimed at leveraging computational algorithms to forecast future events or outcomes based on historical data and observed patterns. At its core, predictive modeling harnesses the power of statistical inference and machine learning techniques to distill actionable insights from complex healthcare datasets, thereby facilitating informed decision-making and risk stratification.

Understanding the Role of Machine Learning in Diagnosis and Prognosis:
Unlike traditional statistical approaches, which often rely on explicit assumptions and predefined models, ML algorithms possess the inherent capacity to autonomously extract meaningful features and iteratively refine predictive models through exposure to empirical data.

By synthesizing disparate sources of clinical data, including electronic health records (EHRs), medical imaging studies, and genetic profiles, ML models facilitate the early detection and classification of diseases, thereby enabling timely interventions and improved patient outcomes.

Similarly, in the domain of prognosis, ML techniques offer

invaluable insights into the trajectory of disease progression and treatment response, thereby empowering healthcare providers to tailor therapeutic regimens to individual patient profiles. Through the analysis of longitudinal patient data and incorporation of temporal dynamics, prognostic models facilitate risk assessment, treatment planning, and patient counseling, fostering a paradigm of personalized medicine and precision healthcare.

Overview of Predictive Modeling Techniques

Predictive modeling encompasses a diverse array of methodologies, ranging from classical statistical techniques to state-of-the-art machine learning algorithms. At its essence, predictive modeling involves the construction of mathematical models that capture the underlying relationships between predictor variables (features) and the target variable of interest (outcome). These models are subsequently trained on historical data to learn patterns and associations, which are then utilized to make predictions on unseen data.

Key predictive modeling techniques include:

Regression Analysis: A classical statistical technique used to model the relationship between a dependent variable and one or more independent variables, regression analysis is widely employed in healthcare for tasks such as predicting patient outcomes, estimating disease risk scores, and assessing treatment efficacy.

Decision Trees: Decision trees represent a non-parametric approach to predictive modeling that recursively partitions the feature space into disjoint regions based on simple decision rules. Decision tree algorithms, such as CART (Classification and Regression Trees) and Random Forests, are adept at handling heterogeneous data types and are particularly well-suited for tasks requiring interpretable models, such as clinical decision support.

Neural Networks: Neural networks constitute a class of deep learning algorithms inspired by the structure and function of the human brain. Characterized by interconnected layers of artificial neurons, neural networks excel at capturing complex, nonlinear relationships within data, making them well-suited for tasks such as medical image analysis, natural language processing, and time-series forecasting.

Support Vector Machines (SVMs): SVMs are a class of supervised learning algorithms that excel at binary classification tasks by identifying an optimal hyperplane that separates data points belonging to different classes. SVMs are widely employed in healthcare for tasks such as disease classification, risk prediction, and outcome modeling.

4.2 Disease Diagnosis Using Machine Learning

Let's consider the scenario of diagnosing diabetic retinopathy, a common complication of diabetes that affects the retina and can lead to vision impairment or blindness if left untreated.

Diagnosing Diabetic Retinopathy:

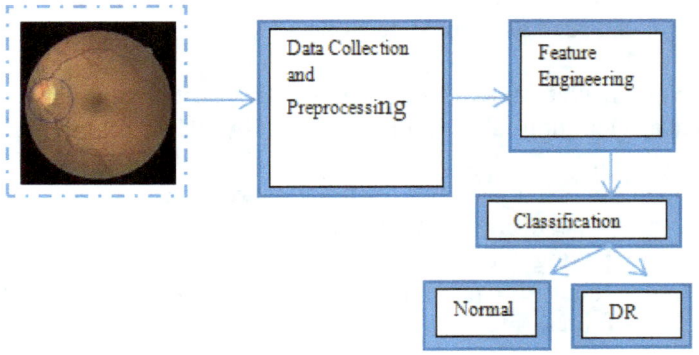

Step 1: Data Collection and Preprocessing
The first step involves gathering a diverse dataset comprising retinal images of patients, annotated with ground truth labels indicating the presence or absence of diabetic retinopathy.

The dataset may also include demographic information such as age, gender, and diabetic history, which can serve as additional features for the diagnostic model. Preprocessing techniques such as image normalization, resizing, and denoising are applied to ensure consistency and quality in the input data.

Step 2: Feature Engineering
Feature engineering plays a crucial role in extracting informative representations from the raw input data. In the context of retinal images, features may include texture patterns, blood vessel morphology, and the presence of lesions characteristic of diabetic retinopathy.

Techniques such as histogram of oriented gradients (HOG), Gabor filters, and local binary patterns (LBP) can be employed to extract relevant features from the retinal images.

Step 3: Model Selection and Training

With the preprocessed data and engineered features in hand, the next step involves selecting an appropriate ML algorithm for training the diagnostic model. Convolutional neural networks (CNNs) have demonstrated exceptional performance in image classification tasks and are well-suited for diagnosing diabetic retinopathy.

The selected CNN architecture is trained on the annotated dataset using techniques such as stochastic gradient descent (SGD) or Adam optimization, with a suitable loss function such as binary cross-entropy.

Step 4: Model Evaluation and Validation

Once the model is trained, it is evaluated using separate validation datasets to assess its performance metrics such as accuracy, precision, recall, and F1-score. Cross-validation techniques such as k-fold cross-validation may be employed to ensure robustness and generalizability of the model across different subsets of the data.

Receiver operating characteristic (ROC) curves and area under the curve (AUC) metrics are utilized to evaluate the model's discrimination ability and balance between sensitivity and specificity.

Step 5: Deployment and Clinical Integration

Upon satisfactory validation, the trained diagnostic model is deployed into clinical practice, where it assists healthcare providers in screening patients for diabetic retinopathy.

Integration with existing electronic health record (EHR) systems enables seamless workflow integration and real-time decision support at the point of care. Continuous monitoring and iterative refinement of the model based on feedback from clinical use cases ensure its ongoing reliability and performance improvement.

By following these steps and leveraging advanced ML

techniques, healthcare practitioners can harness the power of data-driven decision-making to facilitate early detection and intervention for diseases such as diabetic retinopathy, thereby enhancing patient outcomes and reducing healthcare burdens.

4.3 Prognostic Models for Treatment Planning

Prognostic models serve as indispensable tools in healthcare, facilitating informed decision-making regarding treatment strategies and patient management.

Consider a scenario where a team of oncologists aims to develop a prognostic model for predicting the survival outcomes of cancer patients following a specific treatment regimen. The dataset comprises longitudinal patient records encompassing demographic information, clinical characteristics, tumor markers, and treatment modalities. The objective is to construct a robust prognostic model that accurately estimates the probability of survival for individual patients based on these covariates.

Steps in Prognostic Modeling:

1. **Data Preprocessing:** Cleanse the dataset to remove missing values, outliers, and redundant variables. Normalize or standardize numerical features to ensure uniformity in scale. Encode categorical variables into numerical representations using techniques such as one-hot encoding.

2. **Feature Selection and Engineering:** Conduct exploratory data analysis to identify potential predictors of survival outcomes. Utilize domain knowledge and statistical methods (e.g., correlation analysis, feature importance ranking) to select

relevant features. Engineer new features or transformations to enhance predictive power, such as interaction terms or polynomial features.

3. **Model Selection:** Evaluate various machine learning algorithms suited for prognostic modeling, such as Cox proportional hazards regression, random survival forests, or deep survival networks. Employ cross-validation techniques to assess the performance of different models and mitigate overfitting. Choose the model with the optimal balance of predictive accuracy, interpretability, and computational efficiency.

4. **Model Training:** Split the dataset into training and validation sets to facilitate model training and evaluation. Fit the selected prognostic model to the training data, adjusting model parameters to minimize the loss function (e.g., negative log-likelihood for survival models). Incorporate regularization techniques (e.g., L1/L2 regularization) to prevent model overfitting and enhance generalization performance.

5. **Evaluation and Validation:** Assess the performance of the trained model using appropriate evaluation metrics, such as concordance index (C-index) or time-dependent area under the receiver operating characteristic curve (AUC). Validate the model's generalizability on unseen data using holdout validation or k-fold cross-validation. Interpret model coefficients or feature importance scores to glean insights into the underlying prognostic factors.

4.4 Cox Proportional Hazards Regression

Cox proportional hazards regression is a widely used algorithm for modeling the hazard function in survival analysis, allowing for the estimation of covariate effects on survival outcomes while accounting for censoring. The algorithm assumes that the hazard ratio is constant over time, making it suitable for modeling the relative risk of events in prognostic settings.

Implementation of Cox Proportional Hazards regression can be done using pyhton's lifelines library as follows:

i. Importing the CoxPHFitter:

from lifelines import CoxPHFitter

Here, we're importing a specific tool called CoxPHFitter from a Python library called lifelines. This tool helps us apply the Cox Proportional Hazards model to our data.

ii. Instantiating the Model:

cox_model = CoxPHFitter()

We're creating an instance of the CoxPHFitter model. This is like preparing a blank canvas where we'll later fit our data to learn patterns.

iii. Fitting the Model to Data:

cox_model.fit(training_data, duration_col='survival_time', event_col='event_occurred', formula='covariates')

Here, we're training our model using some data (training_data). We tell the model which column represents how long each patient survived (duration_col =

'survival_time') and which column tells us whether the event (like death) happened or not (event_col = 'event_occurred'). Additionally, we specify which variables we want to include in our model using formula='covariates'.

iv. Printing Model Summary:

cox_model.print_summary()

Finally, we print a summary of our model. This summary includes information like coefficients (how strongly each variable affects survival), confidence intervals, and statistical measures of how well the model fits the data. This helps us understand and interpret the results of our analysis.

How Cox Proportional Hazards Works:

- **Identifying Patterns:** Cox Proportional Hazards looks for patterns in our data to understand how different factors (like age, gender, cancer type) relate to survival time.

- **Comparing Hazards:** It compares the "hazard" of one patient group to another. Hazard means the likelihood of an event happening—in this case, death. So, it's like comparing how risky one group is compared to another.

- **Constant Hazard:** It assumes that the ratio of hazards stays constant over time. In simpler terms, it means if one group is twice as likely to experience an event (like death) as another group at the beginning, they'll stay twice as likely throughout the study.

Let's consider a real-life example to predict the survival time of patients with a particular type of cancer after receiving a specific treatment. We'll use hypothetical data in tabular form for illustration :

Patient ID	Age	Gender	Cancer Type	Treatment	Survival Time (months)	Event Occurred
1	55	Male	Lung	Chemotherapy	12	Yes
2	65	Female	Breast	Surgery	36	No
3	70	Male	Colon	Chemotherapy	18	Yes
4	45	Female	Lung	Surgery	24	Yes
5	60	Male	Prostate	Radiation	48	No
.....	.....	.....	.....	.....	....	.....

In this dataset:
- Patient ID: Unique identifier for each patient.
- Age: Age of the patient at diagnosis.
- Gender: Gender of the patient.
- Cancer Type: Type of cancer diagnosed.
- Treatment: Type of treatment received (e.g., chemotherapy, surgery, radiation).
- Survival Time (months): Number of months survived after receiving treatment.
- Event Occurred: Indicates whether the event (e.g., death) occurred during the study period.

Now, let's apply Cox Proportional Hazards Regression to this example:

from lifelines import CoxPHFitter

```python
import pandas as pd

# Load data into a DataFrame
data = pd.read_csv('cancer_data.csv')

# Instantiate the Cox Proportional Hazards model
cox_model = CoxPHFitter()

# Fit the model to the data
cox_model.fit(data, duration_col='Survival Time (months)', event_col='Event Occurred')

# Print model summary
cox_model.print_summary()
```

This code we assumed that the data is stored in a CSV file named '*cancer_data.csv*' and uses the lifelines library in Python. It loads the data into a DataFrame, instantiates the Cox Proportional Hazards model, fits the model to the data specifying the duration column ('Survival Time (months)') and the event column ('Event Occurred'), and finally prints a summary of the model's findings.

The model summary would include information such as coefficients for each variable (age, gender, cancer type, treatment), hazard ratios, confidence intervals, and statistical measures of model fit. These results would help us understand how each factor influences the survival time of cancer patients after treatment.

5. Future Trends and Innovations

Now, Let's explore the trajectory of machine learning in healthcare, elucidating emerging trends and innovative avenues poised to shape the future of medical practice and research. Drawing upon recent research findings and industry developments, we navigate the evolving landscape of healthcare AI, envisioning a horizon replete with transformative possibilities.

5.1 Advancements in Healthcare AI Research

In recent years, the intersection of artificial intelligence (AI) and healthcare has witnessed a proliferation of research endeavors aimed at harnessing the power of machine learning to address the complex challenges inherent in medical diagnosis, treatment, and management. This section delves into the latest advancements in healthcare AI research, highlighting key breakthroughs and their implications for the future of medicine.

Recent Developments
The landscape of healthcare AI research is characterized by a dynamic interplay between technological innovation and clinical application. Recent developments have seen the emergence of novel machine learning algorithms tailored to the specific demands of medical data, including electronic health records (EHRs), medical imaging, and genomic sequences. Deep learning, in particular, has garnered significant attention due to its capacity to automatically learn hierarchical representations from raw data, thereby enabling superior performance in tasks such as image classification, natural language processing, and

predictive modeling.

Researchers have made significant strides in enhancing the interpretability and robustness of AI models in healthcare. Techniques such as attention mechanisms, adversarial training, and uncertainty estimation have been employed to elucidate model predictions, mitigate biases, and quantify uncertainty, thereby fostering trust and transparency among clinicians and patients. Furthermore, efforts to ensure the ethical and responsible deployment of AI in healthcare have led to the formulation of guidelines and frameworks emphasizing principles such as fairness, accountability, and transparency (FAIR).

Integration of Multi-modal Data

The advent of precision medicine has underscored the importance of integrating multi-modal data sources, including genomic, clinical, imaging, and environmental data, to elucidate the interplay between genetic predispositions, environmental exposures, and disease phenotypes. Machine learning techniques such as multi-task learning, transfer learning, and federated learning have enabled the integration of heterogeneous data modalities, facilitating holistic assessments of patient health and disease trajectories.

For instance, researchers have leveraged deep learning algorithms to analyze multi-modal medical imaging data, including magnetic resonance imaging (MRI), computed tomography (CT), and positron emission tomography (PET) scans, to improve diagnostic accuracy and treatment planning in conditions ranging from neurodegenerative diseases to cancer. By extracting high-dimensional features and spatial relationships from imaging data, AI models can discern subtle patterns indicative of disease presence, progression, thereby empowering clinicians with actionable insights and decision support tools.

Explainable AI

As AI models become increasingly pervasive in clinical settings, the need for explainable and interpretable AI (XAI) has become paramount to engendering trust and fostering adoption among healthcare stakeholders. Explainability refers to the ability of AI models to provide transparent and comprehensible rationales for their predictions or decisions, thereby enabling clinicians to understand the underlying mechanisms driving model outputs and to validate their clinical relevance.

Recent advancements in XAI techniques have encompassed a spectrum of approaches, including model-agnostic methods such as LIME (Local Interpretable Model-agnostic Explanations) and SHAP (SHapley Additive exPlanations), as well as model-specific techniques such as attention mechanisms and saliency maps. These techniques aim to elucidate the contributions of individual features or data points to model predictions, thereby facilitating the identification of clinically relevant biomarkers, risk factors, and treatment responses.

5.2 Integration of ML into Medical Practice

In recent years, the integration of machine learning into medical practice has garnered significant momentum, with AI-driven tools and technologies reshaping clinical workflows, enhancing diagnostic accuracy, and improving patient outcomes. This section explores the burgeoning role of machine learning in medical practice and its transformative impact on healthcare delivery.

Clinical Decision Support Systems (CDSS)
Clinical decision support systems (CDSS) powered by machine learning algorithms have emerged as indispensable tools for augmenting clinical decision-making processes. These systems leverage patient data, including electronic health records (EHRs), medical imaging, and laboratory results, to provide clinicians with real-time insights and recommendations tailored to individual patient profiles.

Machine learning algorithms, ranging from traditional statistical models to deep learning architectures, are employed to analyze vast amounts of patient data and extract clinically relevant information, such as disease risk scores, treatment recommendations, and prognostic outcomes. By integrating AI-driven CDSS into clinical workflows, healthcare providers can leverage evidence-based guidelines, clinical protocols, and predictive analytics to optimize treatment decisions, reduce diagnostic errors, and improve patient safety.

Precision Health Initiatives
Precision health initiatives aim to harness the power of data-driven technologies, including machine learning and genomic sequencing, to tailor medical interventions according to individual patient characteristics, preferences, and genetic predispositions. By leveraging large-scale datasets encompassing genomic, clinical, and environmental data, researchers can identify biomarkers, genetic variants, and phenotypic traits associated with disease susceptibility, progression, and treatment response. Machine learning algorithms play a pivotal role in deciphering the intricate interplay between genetic, environmental, and lifestyle factors shaping individual health trajectories. Techniques such as polygenic risk scoring, genome-wide association studies (GWAS), and deep learning-based phenotype prediction models enable the identification of disease subtypes, stratification of patient populations, and personalized treatment recommendations. Through precision health initiatives, healthcare providers can move beyond a one-size-fits-all approach to healthcare delivery, towards a paradigm of tailored interventions and preventive strategies tailored to the unique needs of each patient.

Telemedicine and Remote Monitoring
The advent of telemedicine and remote monitoring technologies facilitated by machine learning has revolutionized the delivery of healthcare services, particularly in remote or underserved communities.

Telemedicine platforms enable clinicians to conduct virtual consultations, diagnose conditions, and prescribe treatments remotely, thereby overcoming geographical barriers and enhancing access to care.

Machine learning algorithms are employed to analyze patient-generated data, including wearable sensor data, physiological signals, and self-reported symptoms, to monitor disease progression, detect early warning signs, and optimize treatment regimens. By leveraging AI-driven predictive analytics and remote monitoring technologies, healthcare providers can proactively intervene to prevent complications, optimize resource allocation, and improve patient outcomes, while empowering patients to actively participate in their own care.

5.3 Future of Healthcare Technology

As we look ahead to the future of healthcare technology, the convergence of artificial intelligence (AI) and advanced technologies holds immense promise for revolutionizing medical practice, research, and patient care. This section explores emerging trends and innovations poised to shape the trajectory of healthcare technology in the years to come.

AI-driven Drug Discovery

The traditional drug discovery process is characterized by high costs, lengthy timelines, and a high rate of attrition. However, the advent of AI-driven drug discovery platforms offers a paradigm shift in pharmaceutical research, accelerating the drug development pipeline and enhancing therapeutic efficacy. Machine learning algorithms are employed to analyze large-scale biomedical datasets, including chemical libraries, genomic sequences, and clinical trial data, to identify novel drug targets, predict drug-drug interactions, and optimize drug candidates.

By leveraging AI-driven predictive modeling, researchers can expedite the identification of promising drug candidates, prioritize lead compounds for further validation, and optimize therapeutic regimens tailored to individual patient profiles. Furthermore, machine learning algorithms enable the identification of biomarkers, genetic variants, and molecular signatures associated with drug response and resistance, facilitating the development of personalized treatment strategies and precision therapeutics. Through AI-driven drug discovery, we can usher in an era of faster, safer, and more efficacious treatments for a wide range of diseases, from cancer to rare genetic disorders.

Augmented Reality (AR) and Virtual Reality (VR) in Healthcare

Augmented reality (AR) and virtual reality (VR) technologies hold transformative potential for enhancing medical education, surgical training, and patient rehabilitation. By overlaying digital information onto the physical environment, AR enables clinicians to visualize anatomical structures, medical images, and procedural guidance in real time, enhancing situational awareness and decision-making during surgical procedures. Virtual reality (VR) simulations offer immersive training environments where clinicians can practice surgical techniques, refine procedural skills, and navigate complex anatomical scenarios in a risk-free setting.

Machine learning algorithms are integrated with AR and VR technologies to personalize training simulations, adapt to individual learning styles, and provide real-time feedback on performance. Furthermore, AI-driven analytics enable the analysis of user interactions, procedural errors, and skill acquisition metrics, facilitating continuous improvement and optimization of training protocols. Through the integration of AI with AR and VR technologies, we can empower healthcare professionals with the knowledge, skills, and confidence necessary to deliver high-quality care and improve patient outcomes.

Global Health Equity

In the pursuit of global health equity, artificial intelligence (AI) holds promise as a transformative tool for addressing healthcare disparities and advancing access to quality care for underserved populations. By leveraging AI-driven technologies, such as mobile health applications, telemedicine platforms, and community-based interventions, we can overcome geographical barriers, bridge the digital divide, and extend the reach of healthcare services to remote or resource-limited settings.

Machine learning algorithms are employed to analyze population-level health data, identify vulnerable populations, and predict disease outbreaks, enabling proactive interventions and targeted public health strategies. Furthermore, AI-driven diagnostic tools and point-of-care devices empower frontline healthcare workers with the ability to accurately diagnose and treat common diseases, monitor disease progression, and track treatment outcomes in real time. Through collaborative efforts and innovative solutions, we can harness the power of AI to achieve universal health coverage, promote health equity, and improve the well-being of communities worldwide.

Conclusion

In wrapping up our journey through the world of "Healing with Data: Machine Learning in Healthcare," it's clear that the future of medicine is filled with exciting possibilities. We've seen how artificial intelligence (AI) and machine learning are not just buzzwords, but powerful tools that can revolutionize how we diagnose illnesses, treat patients, and even discover new medicines.

Throughout this book, we've explored how AI algorithms can analyze massive amounts of medical data—from X-rays and MRIs to genetic information—and help doctors make better decisions. We've learned how these technologies can predict diseases, personalize treatments, and even assist in surgeries.

But it's not just about the technology—it's about what it means for people. By harnessing the power of AI, we can make healthcare more accessible and equitable for everyone, regardless of where they live or how much money they have. Telemedicine and remote monitoring, powered by AI, can bring medical expertise to remote areas. And by understanding each patient's unique needs through precision medicine, we can deliver treatments that are tailored specifically to them.

As we look to the future, it's important to remember that while AI is powerful, it's not a magic bullet. We must continue to ensure that these technologies are used ethically, responsibly, and with the utmost respect for patient privacy. We must also address issues of bias and fairness to ensure that AI benefits everyone equally.

In closing, "Healing with Data" has been a journey into the exciting intersection of healthcare and technology. As we move forward, let's continue to explore, innovate, and collaborate to harness the full potential of AI in healthcare, making the world a healthier and happier place for all.

www.ingramcontent.com/pod-product-compliance
Lightning Source LLC
Chambersburg PA
CBHW050014230526
45470CB00003B/969